Cat Wisdom

60 Great Lessons You Can Learn from a Cat

NEIL SOMERVILLE

Thorsons

Thorsons
An imprint of HarperCollins*Publishers*
1 London Bridge Street
London SE1 9GF

www.harpercollins.co.uk

First published by Thorsons 2017

1 3 5 7 9 10 8 6 4 2

A catalogue record of this book is available from the British Library

ISBN 978-0-00-825275-5

Printed and bound in Great Britain by CPI Group (UK) Ltd, Croydon

Cat Wisdom

This book is dedicated to cats.
To Roly, Smartie, Lily and Harry, and
to your cat or a cat you may know.

Contents

Introduction	2
Persistence	4
Gratitude	6
At Ease	8
Quiet Observation	10
Special Greetings	12
The Importance of Giving	14
Discover	16
Taking Care	18
Keeping Your Head Down	20
Letting Off Steam	22
Sleep	24
Buying Time	26
Stretching	28
Reward	30
Looking Good	32
Courage	34

Making Do 36
Language of the Eyes 38
Surroundings 40
Awareness 42
Reflecting 44
Downtime 46
Scent 48
Decision Time 50
Adapt 52
Responsibility 54
Tell-tail Signs 56
Worth the Wait 58
Taking Stock 60
Friendship 62
Value the Now 64
Nine Lives 66
Favourite Places 68

Venture Hopefully 70
Exercise 72
Listening Out 74
Seize the Moment 76
Sweet Dreams 78
Stress 80
Private Moments 82
Going for Best 84
Being Attentive 86
Beware the Stare 88
You Never Know 90
Sensing 92
Moving On 94
Routine 96
Enthusiasm 98
On Making It Clear 100

Standing Up 102
Win-win 104
Poised to Do Well 106
Self-sufficiency 108
Knowing Where 110
Look Before You Leap 112
Getting to Know You 114
Contentment 116
Make Your Mark 118
Never Too Late 120
Body Clock 122
Final Thoughts 124
List of Lessons in Alphabetical Order 127
About the Author 129
Acknowledgements 131

Introduction

For so many of us, cats have a special place in our lives. With their affection, grace, companionship and personality, cats can give us great joy and, in so many ways, they can also teach us – and remind us – of helpful truths. Whether through their traits, the way they so effectively set about things or through our observing and spending time with them, cats demonstrate that they have great wisdom.

In the pages that follow are sixty lessons inspired by the cat.

Enjoy them, think about them, and I very much hope that they may inspire and enrich you in some way.

Persistence

'Cats seem to go on the principle that it never does any harm to ask for what you want.'

JOSEPH WOOD KRUTCH

When a cat wants feeding it will leave you in no doubt about what it wants. Whether it's meowing loudly, standing in your way or looking pleadingly at the food cupboard, it is virtually impossible to ignore a determined cat. And the cat will keep on at you until you give in.

The cat is well aware of the importance of persisting, and persistence often pays. If there is something you want, don't keep it to yourself and risk disappointment; be like the cat. Be forthcoming and ask. And ask again if necessary. It is through asking that you are more likely to receive.

Gratitude

'To please himself only the cat purrs.'

IRISH PROVERB

The cat, the master of the good life, enjoys being pampered. And whether this is by being stroked or given attention, the cat indicates its pleasure with a deep and rhythmic purr. It is a cat's way of saying 'Thank you' and, because we know it is grateful, we tend to carry on with the pampering even longer.

By showing appreciation of what others do, it will please those concerned, as well as make them more likely to respond positively to you again.

Be like the cat: appreciate and *acknowledge* the favours and kindnesses given to you.

At Ease

'No day is so bad it can't be fixed with a nap.'

CARRIE SNOW

One of the cat's delights is in taking a catnap. Whether it's for a few moments beside a warm fire or in a sunny spot in the garden, the cat loves to curl up and enjoy a snooze.

This is something we too can benefit from. A few moments' rest during the day can do us so much good. Either find somewhere quiet where you can close your eyes for a few minutes and get away from the noise and bustle around you, or, if you are able, lie down for a short time during the day. Such a break will leave you feeling better, refreshed *and* often more productive too.

Be like a cat and enjoy the benefits of an occasional catnap.

Quiet Observation

'Cats are mysterious kind of folk – there is more passing in their minds than we are aware of.'

SIR WALTER SCOTT

Whether from a branch of a tree, a flat roof or some other location, the cat has a great talent for selecting vantage points where it can observe what is going on around it while remaining unnoticed.

Quiet observation is something we too can use to our advantage. Often our lives are so busy that we rarely stop to take a look at where we are, what we are doing or what our general position and purpose is. But if we adopt the cat's practice and quietly observe and evaluate what is going on around us as well as the things we do, we could certainly notice more, and probably come up with more ideas and possibilities we can build on. Be like the cat and take time to quietly observe; your understanding and awareness will be so much greater as a result.

Special Greetings

'There are few things in life more heartwarming than to be welcomed by a cat.'

TAY HOHOFF

Many a contented cat, happy in your presence, will roll over and lie on its back, belly-up. Sometimes it will even allow you to stroke its soft, furry underside, although should you touch a tickly spot, beware! For the cat to lie on its back in such a way is a sign of trust and friendliness. It is a special greeting, reserved for a select few.

We can follow the cat's example and make a special effort when meeting friends. Show a warmth and genuine pleasure at seeing them, rather than giving just a casual 'Hiya'. Make it apparent it *is* good to see them. Smile, show enthusiasm, radiate warmth, and by making the effort, your presence and relations with those around you will often become richer *and* more meaningful.

The Importance of Giving

'Cats can be very funny, and have the oddest ways of showing they're glad to see you. Rudimace always peed in our shoes.'

W. H. AUDEN

For a cat to present you with its prey is an honour. Whether this is laying a dead mouse on your doorstep or bringing it indoors, it's a cat's way of showing love, care and trust. It is akin to a mother cat sharing food with her kittens, and your cat, with its present, wants to show you its love and that you won't starve or go without.

Admittedly, a dead mouse is a present you may not necessarily appreciate, but present-giving is a sign that you care about and are keen to please another. And when the present is unexpected and well thought out, it will often mean that much more. Just like a cat, give an occasional present to those who are important and special to you – only choose something more appropriate!

Discover

'Watch a cat when it enters a room for the first time. It searches and smells about, it is not quiet for a moment, it trusts nothing until it has examined and made acquaintance with everything.'

JEAN-JACQUES ROUSSEAU

Whenever anything new is introduced to the home, the cat invariably wants to find out more. It looks, it watches, it sniffs. If it is food and shopping bags, the cat will be quick to detect any wafts of fresh fish or other cat delicacy. Or if it is soft furnishings, again the cat will be keen to investigate and possibly view it as a place to nestle, sleep or possibly scratch. Even a new washing machine could come in for a thorough sniffing before the cat decides that washing machines are not really a cat thing.

By noticing and investigating, the cat learns, discovers and sometimes benefits. And so it is with us. Rather than close our minds to anything new that comes along, we should at least try to find out more. It's by being curious, by asking, by taking a look and having an open mind that we learn, grow and ultimately make more of ourselves.

Like a cat, enjoy the thrill of discovery.

Taking Care

'It is in the nature of cats to do a certain amount of unescorted roaming.'

ADLAI E. STEVENSON

Often, when strolling across a garden, the cat will stop and chew a few blades of grass. And although grass may seem a curious choice for a cat and carnivore, it is the cat's way of getting folic acid, an important vitamin for its digestion.

We should follow the cat's example and make sure we eat and seek out food that is good for us. Fruit and vegetables, fibre, grains and pulses can all help keep our system in good order, and if we are to live and enjoy the good life like the cat, looking after ourselves and watching what we eat is a key and sensible priority.

Keeping Your Head Down

'The cat seldom interferes with other people's rights. His intelligence keeps him from doing many of the fool things that complicate life.'

CARL VAN VECHTEN

When someone raises their voice, many a cat will turn tail and move to another area. And if the cat cannot go to another room, it could well crouch down, keep still and hope it will go unnoticed.

Cats have a great dislike of tension and do their best to avoid hostile and threatening situations. Loud voices or shouting, in particular, make many wary. And this is a lesson we too can learn from. If we detect that someone is about to lose their temper or a situation is becoming fraught, we can often, if possible and practical, save ourselves from heated exchanges and allow the situation to cool down by getting out of the way and keeping a low profile.

At times it can be politic to 'keep your head down'.

Letting Off Steam

'A kitten is chiefly remarkable for rushing about like mad at nothing whatever, and generally stopping before it gets there.'

AGNES REPPLIER

One curious behaviour of many a cat is to suddenly go for a mad dash around the room, first hurtling in one direction, then in another, before finally coming to a halt and returning to its more normal and placid state.

A reason cats do this is to help them release pent-up energy, particularly if they've been inactive for any length of time. In our own lives, there can be days when we don't get much physical activity and our energy levels build up, sometimes causing restlessness and problems getting to sleep. Like cats, we too need to give ourselves the chance to release some of our excess energy. Going for a walk, exercising or carrying out any physical activity that requires a bit of exertion can often leave you feeling more relaxed and less stressed.

Like a cat, when you have too much energy, it can be helpful to let off steam.

Sleep

'You cannot look at a sleeping cat and feel tense.'

JANE PAULEY

Each day a cat will spend about twelve to sixteen hours in blissful slumber. While asleep, the cat not only rests, but also builds up energy, ready for the rigours ahead.

Although our own sleep needs are different, like the cat we must not forget the value of sleep. Sleep allows us to unwind, assimilate thoughts and digest happenings, and prepares us mentally and physically for the day ahead. It's essential to our well-being, and yet with the demands of modern-day life it can be tempting to cut back on sleep and keep long hours. Depriving ourselves of the rest our bodies need, however, can start to have a debilitating effect. Like a cat, we need to recognise the importance of sleep and how it not only helps our general health and mood, but also allows us to be more alert, productive and energised.

Do not ignore the value of sleep – something a cat knows all too well.

Buying Time

'Intelligence in the cat is underrated.'

LOUIS WAIN

Many a cat has pleaded for food and then, when it has been placed in its bowl, thought better of it and walked away. Occasionally the cat will relent and return for a nibble, but sometimes it will hold out until it's served something better. The cat knows that it is not always best to take what is offered first time, but rather to check what else is available and then decide.

We too can use this waiting strategy to good effect. If something is offered, rather than accept it without quibble, take time to think about what is actually being offered, whether it's what you want, what it entails and whether it could be bettered. Sometimes it can, sometimes it can't, but as the cat knows, it can sometimes be worth buying that extra time and enjoying the benefits that patience and self-control can bring.

Stretching

'If stretching were wealth, the cat would be rich.'

AFRICAN PROVERB

When a cat rouses after a rest or nap, it often enjoys a good stretch. This not only helps to get its muscles moving again and improve circulation; it also readies the cat for any activities ahead.

Stretching can be highly beneficial, and with many of us sitting still for long periods during the day, an occasional stretch can help reinvigorate us, provide relief for stiff muscles and improve posture, flexibility and well-being. Cats stretch for a good reason and there is much to be gained from incorporating this practice into your day and finding out more about routines and exercises that could be beneficial.

‘Cats can be cooperative when something feels good, which, to a cat, is the way everything is supposed to feel as much of the time as possible.’

ROGER A. CARAS

A cat, like all of us, enjoys a good reward. If being trained to go through a cat flap, a tasty morsel could be the temptation it needs. Similarly, knowing that, when called, it's likely to receive some food, the cat will be keen to comply. It's worth making the effort for the reward.

So it is with us. The prospect of an inducement can help spur us on, and when there is something you want to achieve, it is worth keeping in mind the end result and the rewards that can come from your actions. For a cat, food can be inducement enough, while for us the potential rewards that can come from good and determined effort can be almost limitless.

Looking Good

'Cats are living adornments.'

EDWIN LENT

A cat spends a long time grooming, and for good reason. It helps to keep its fur clean and also maintains its protection against the elements. In addition, if recently stroked or handled, a good lick will help to smooth its fur as well as replace human odour with its own. The cat takes great care and pride in itself, and benefits from the process.

This applies to us as well. By paying attention to ourselves – our hygiene, appearance and clothes – it not only helps to make us feel better physically, but also leaves us more confident in ourselves. As with the cat, a thorough groom can help us look good, but it can also be good for our body *and* our self-esteem.

As has often been said, 'Appearance counts,' and looking good does good in so many ways.

Courage

'Way down deep, we're all motivated by the same urges. Cats have the courage to live by them.'

JIM DAVIS

There are many accounts of cats who have become separated from their home and have trekked vast distances to return.

To start on such a quest requires great courage and the cat must know that if it is ever to reach its destination, it must face many risks.

In our own lives there can also be times when we are desperate to accomplish something and, like the cat, must venture far out of our comfort zone to achieve it. Yes, it will take courage, but if we are to stand any chance of getting what we want, journey we must.

The courage of some cats has been truly remarkable, and in our own quests we too need to draw on our reserves and harness the extraordinary abilities and strengths that lie within. As Virgil noted, 'Fortune favours the bold,' and it is the bold, and those who get on and do, who most often reach their destination.

Cats can be courageous. So can we.

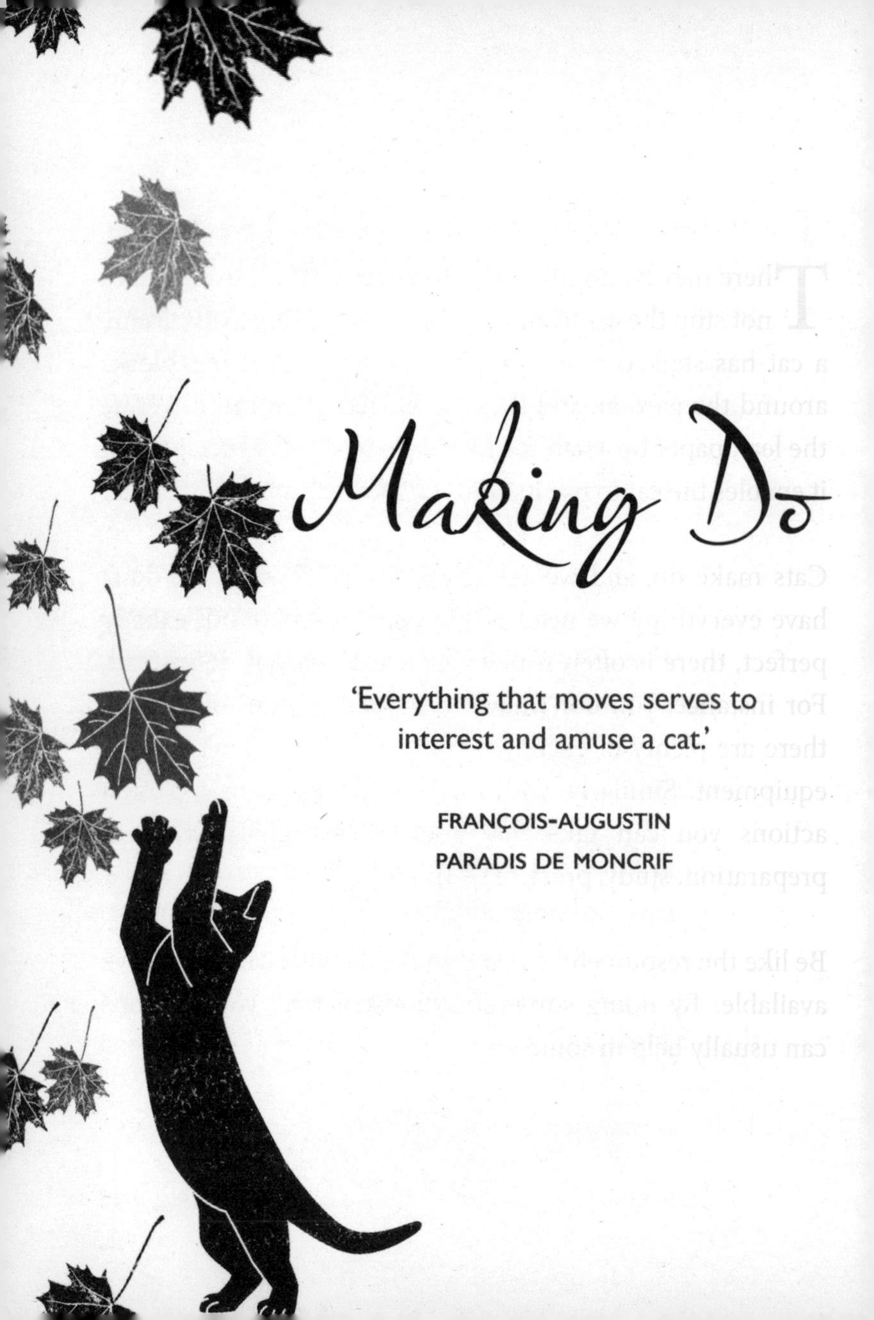

Making Do

'Everything that moves serves to interest and amuse a cat.'

FRANÇOIS-AUGUSTIN PARADIS DE MONCRIF

There may be no mice or other prey about, but this does not stop the cat from practising its hunting skills. Many a cat has stalked a leaf or scrap of paper as it has blown around the garden, and then, when ready, pounced. While the leaf, paper or whatever may only be a substitute, at least it enables the cat to use its skills and keep them finely tuned.

Cats make do, and we can do the same. Even if we don't have everything we need or the conditions are not exactly perfect, there is often something useful we can do instead. For instance, you don't have to go to the gym to keep fit; there are plenty of exercises you can do that don't require equipment. Similarly, with other activities there are often actions you can take that can be helpful, whether in preparation, study, practice or thought.

Be like the resourceful cat and make do with what you have available. By doing something constructive, your actions can usually help in some way.

Language of the Eyes

'It is in their eyes that their magic resides.'

ARTHUR SYMONS

In relation to its small head, a cat's eyes are very large. Eyes and vision are especially important to the cat, not only for hunting and seeing well in different lights (which it can do thanks to its vertical pupils), but also as a way to communicate.

When a cat has its eyes wide open, it's a sign that it is alert and, with a fixed stare, aggressive. If a cat half-closes its eyes, it is signalling that it is more at ease.

Eye language can be very indicative, particularly of mood and inner feelings – not for nothing have eyes been called the 'windows of the soul'. Our own eyes can say a great deal too. A widening of the eyes can show appeal and interest, while looking up and to the left can be an indication that we are reminiscing and drawing on memories. Conversely, looking upwards and to the right suggests we are thinking more creatively. But beware – looking down and avoiding eye contact can be a sign of disinterest.

Cats use their eyes to communicate, and although we may not always be consciously aware, so do we. In your dealings and interactions, observe the eyes of others; it can often help your understanding of how – and even what – they may be thinking.

Surroundings

'When she walked ... she stretched out long and thin like a little tiger, and held her head high to look over the grass as if she were treading the jungle.'

SARAH ORNE JEWETT

Territory and surroundings are especially important to cats and they make sure they know them well. For them, territory can be a source of food, as well as a place of safety.

Their core territory is often the home, but cats also have their own territories outside, which they scent mark and use to their advantage.

However, while we, like the cat, attach much importance to our home, we do not always make the most of what is available in our immediate area. In many a neighbourhood and town there are resources and facilities we can use, as well as entertainments to enjoy.

For the cat, its surroundings offer a veritable feast of things to notice and do, and for us too the area around us can hold many delights; more than you may realise. Be like the cat: get to know more about the area where you live and discover the joys and benefits it offers.

Awareness

'Cats can read your character better than a $50 an hour psychiatrist.'

PAUL GALLICO

When you are off colour and not at your best, many a cat will come and sit with you or lie quietly by your side. A cat seems very attuned to how we feel and offers not only its presence, but also comfort when we need it. And in doing so it can be very therapeutic.

This is a valuable lesson that we can learn from the cat. If someone you know is unwell or not feeling special, show them you care, either by offering to help or asking after them. Your actions and words can be much appreciated – more so than you may realise – and lift the spirits of someone who is not feeling great.

This also applies in our daily dealings with others; thoughtfulness and well-meant enquiries are often of inestimable value.

Be like the cat: notice and show understanding. A simple gesture can make a surprising difference.

Reflecting

'The cat dwells within the circle of her own secret thoughts.'

AGNES REPPLIER

At times during the day, many a cat will sit, quiet and still, lost in thought. We may not have the privilege of knowing what it's thinking, but as it sits, its body at rest, the cat seems very content.

We too can learn and benefit from this meditative process, particularly allowing ourselves a time to be still and to do our own thinking. A moment of stillness can be a great way to nourish body and mind, giving the chance for ideas to occur and thoughts to fall into place, as well as bringing a feeling of calm.

As Pythagoras declared, 'Learn to be silent. Let your quiet mind listen and absorb.'

You'll feel so much better as a result.

Downtime

'Cats are magical … the more you pet them the longer you both live.'

ANONYMOUS

A cat, with its presence, brings joy to many a home, and this joy can be made all the greater by sharing and spending time together.

Playing with a cat can not only increase the bond that may exist between you; it may also be mutually beneficial. For the cat, playing can be mentally stimulating, it can burn off some of its excess energy (sometimes a cause of behavioural problems) and it can also be good for confidence, especially by encouraging it to engage and interact with others.

And for us, playing can be fun, giving us a chance to relax, unwind and enjoy some valuable downtime – something which is important in our often busy lives.

Cats can teach us many things, but sharing time together does great good, usually leaving the cat purring and you smiling.

We all need some downtime, and when we get it, we feel all the better for it.

Scent

'Catnip is vodka and whisky to most cats.'

CARL VAN VECHTEN

Sense of smell is very important to cats, so much so that they have about 200 million scent receptors in their noses, compared to our 5 million.

Cats use scent as a means of identification; when they rub against you, they are transferring some of their scent to you so that you become part of their social group. They also use scent markings to communicate to other cats. And should they happen to smell catnip, some can almost be sent into a state of ecstasy, such is its effect.

Just as cats attach much importance to scent, we too can use it to our advantage. Perfumes or aftershaves can be pleasant as well as mood enhancing, and the fragrance of flowers, newly baked bread and freshly ground coffee, along with other tempting smells, are something many people enjoy and appreciate. Aromas can also be used to boost our general well-being; lemon and peppermint are considered helpful for mental alertness, while lavender promotes a feeling of calm and, for some, aids a restful night's sleep.

Cats pay much attention to scent and, in our own way, we can use it too to make our surroundings all the more pleasant and conducive, including for rest and relaxation.

Decision Time

'I have studied many philosophers and many cats.
The wisdom of cats is infinitely superior.'

HIPPOLYTE TAINE

If a cat has the chance to go outside and then, just as it's about to venture out, discovers that it's wet and windy and not so inviting, it will wag its tail as it decides what to do. Similarly, if there is a choice between a nap in front of a warm fire or a longer sleep on a comfortable bed, the cat is likely to wag its tail as it makes up its mind. This tail-wagging – first one way and then the other – shows the cat is thinking and torn between two choices. It's also a signal from the cat to 'leave me alone and let me think this through by myself'.

Although we don't have (or need) a tail to wag, we should tell others when we need some thinking time and ask for quiet, or go to a place where we can think, undisturbed. Important decisions *do* need time and quiet reflection, and when you have this, your decisions will often be stronger and better as a result.

Adapt

'The really great thing about cats is their endless variety. One can pick a cat to fit almost any kind of decor, colour, scheme, income, personality, mood. But under the fur, whatever colour it may be, there still lies, essentially unchanged, one of the world's free souls.'

ERIC GURNEY

One of the reasons cats are such popular pets is because they fit in well to many a home. Quiet and private, yet affectionate, a cat successfully works its way into our lives and we are all the better for it. But when it ventures out, the cat reverts to its own feline world and ways. Cats are very good at adapting. Similarly, feral cats have also shown themselves adept at adjusting to modern-day life, with their numbers increasing.

Just as cats have shown how successful they are at adapting, so must we embrace adaptability if we are to thrive. We live in fast-moving times, with new systems, procedures and technology all having an impact. Rather than resist, it is important we keep up with developments and adjust accordingly. To remain wedded to old procedures risks being left behind, so as individuals we should learn from the way the cat has so successfully adapted to the conditions of the time. By doing likewise, we can make far more of the opportunities the present day affords.

Responsibility

'Even a cat is a lion in her own lair.'

INDIAN PROVERB

Although the cat enjoys company, whether that's living with its owner or meeting and playing with other cats, it also needs a certain amount of independence. Particularly when hunting, the cat goes it alone. This is the cat's own preserve and responsibility, and if in the wild, its very existence and survival would be at stake.

This sense of responsibility applies to us as well. What happens in our own lives is the result of what we do and how we use our skills, abilities and chances. But unlike the cat, for us it can sometimes be easy to look for excuses or to 'pass the buck'. Be like the cat: take responsibility for yourself and your actions, for so much does, in effect, hinge upon what *you* do. The cat may enjoy the good life and take advantage of chances provided by others, but it also knows that it has to fend for itself. The cat is never one to shirk responsibility; neither should we.

Tell-tail Signs

'I saw the most beautiful cat today. It was sitting by the side of the road, its two front feet neatly and graciously together. Then it gravely swished around its tail to completely encircle itself. It was so fit and beautifully neat, that gesture, and so self-satisfied, so complacent.'

ANNE MORROW LINDBERGH

The cat uses its body to telling effect. Whether in how it holds its tail, the position of its ears (which it can move quite considerably) or its general stance, it is able to convey its mood and send messages to others. For instance, with ears and tail upright, that is generally a positive signal, but with ears pulled downwards and tail bristled the cat is showing itself to be anxious. And should you see a cat running towards you with its upright tail quivering, that is usually a sign it is really pleased to see you.

Just as cats use their body so effectively to communicate, so can we, and in so doing we need to be aware of our own body signals. To stand with folded arms can, for instance, form a barrier between you and another, whereas if someone adopts (and mirrors) a similar posture to your own, that is generally a sign of accord and agreement. Leaning forward in the direction of another can also signify interest, as can an encouraging nod of the head.

Cats are very adept at getting their feelings across using their bodies, and by being aware of our own gestures and postures, we too can use body language to convey our feelings and communicate with others.

Worth the Wait

'A cat knows how to anticipate.'

ROGER A. CARAS

When a cat is hunting and discovers a rodent's burrow it will stay and watch it intently, ready to pounce the moment anything appears. This stalking and watching can last a long time, but the cat knows that if its hunt is to be successful, it must play the waiting game.

In our own lives, some of the things we want may not be immediately available, but if they really matter, then we must also be prepared to be patient and know that our objective, and the benefits it can bring, are worth the wait. As Jean-Jacques Rousseau declared, 'Patience is bitter, but its fruit is sweet.'

Like a cat, be prepared to wait, but be ready to pounce the moment the opportunity appears.

Taking Stock

'Of all animals, the cat alone attains to the contemplative life.'

ANDREW LANG

On many a prowl a cat will stop in its tracks and stand still for a moment, reassessing its situation, taking in what is around it and listening. In our own lives, taking a pause can be similarly valuable. It allows us to collect our thoughts, take note of developments and review what we are doing and our approach. Rather than conducting so much of our lives at a heady pace, it pays to take stock every so often, assess conditions and so be better prepared for our next move, just like the cat.

Friendship

'But once gain his confidence, and he is a friend for life. He shares your hours of work, of solitude, of melancholy. He spends whole evenings on your knee, purring and dozing, content with your silence, and spurning for your sake the society of his kind.'

THÉOPHILE GAUTIER

During the day, many a cat will offer friendship and companionship. This could be snuggling down on someone's lap, welcoming a family member home or joining in when suitable opportunities present themselves (especially for play), as well as being part of a family. By offering friendship, the cat has become a friend to many.

Our own lives too can be made so much richer by offering friendship and reaching out to others. If alone, see what is available in your community by way of activities you could join in with to get to meet others. And with existing friends, ask after them, show support, encourage them and be prepared to listen and *be a friend*. As cats show, friendship is an important part of enjoying the good life and can add much value to our own.

As the philosopher Thomas Aquinas wrote, 'There is nothing on this earth more to be prized than true friendship.'

Value the Now

'The reason domestic pets are so lovable and so helpful to us is because they enjoy, quietly and placidly, the present moment.'

ARTHUR SCHOPENHAUER

Cats live in the present and invariably make the most of their situation. If it is warm and inviting outside and the cat is able, it will go out and enjoy the good weather. Similarly, if cold and wet, curling up on a dry, warm bed will do very nicely. And if there is good company about, the cat will be keen to join in.

Cats value the now and so often make the most of their situation. While in our own lives there can be numerous pressures and umpteen things to do, there is still much to appreciate in the now. Whether it's time spent with loved ones or your pet(s), or current activities, projects and interests, there is much to be thankful for and good to be done.

As the cat so frequently shows, we should enjoy and make much of now. Now is *our* time, and a time to value.

Nine Lives

'It has been the providence of Nature to give
this creature nine lives instead of one.'

PANCHATANTRA

It has often been said that a cat has nine lives, because in its lifetime a cat will usually get into many scrapes or predicaments. Sometimes this can be caused by its own curiosity or adventurous nature, but in many cases the cat survives unscathed and recovers well from its ordeal.

In our own lives, there will undoubtedly be difficult times to face too. But, like the cat, we are incredibly resilient. Out of adversity we learn, we adjust and sometimes make fresh starts. Challenges provide lessons; they make us who we are and are there to be overcome.

We may not have the nine lives of the cat – and hopefully we won't need them – but, with our inner strengths and capabilities enabling us to triumph over adversity, we can prove ourselves to be just as adept at surviving as the cat.

Favourite Places

'Wherever a cat sits, there shall happiness be found.'

STANLEY SPENCER

Many a cat has its favourite place – under the shade of a particular bush, on a certain windowsill, on a garage roof or the branch of a tree – and it will often gravitate there, feeling safe and content.

A favourite place can make us feel good too. Whether because of its atmosphere, serenity or beauty, or because it allows us to be ourselves, time spent there can be restorative and relaxing. For some, that place could be a shed, a 'man cave', a garden or beauty spot, or just somewhere we regard as special, but wherever it is, such a place is good for the psyche.

Cats enjoy their favourite places ... and we should follow their example and treasure and enjoy ours. A place for you just to be.

Venture Hopefully

'In the middle of a world that has always been a bit mad, the cat walks with confidence.'

ROSANNE AMBERSON

When cats go on the prowl, they go in a state of expectation. Maybe they'll see and catch some prey, enjoy a stroke or titbit from a friendly neighbour or find something to play with. However, the cat prowls for a purpose, and because it's hopeful that something will turn up, it usually does.

This can happen in our own lives too. If we are optimistic, expecting something good to happen or perhaps hoping for an idea or opportunity to occur, then it often will. As J. B. Morgan and Ewing T. Webb noted, 'When you expect things to happen – strangely enough – they do happen.' Be like the cat: as you set about your activities, venture in hope.

Exercise

'For me, one of the pleasures of cats' company is their devotion to bodily comfort.'

SIR COMPTON MACKENZIE

A cat is a hunter, and to be successful it needs to keep nimble and fit. During the day many a cat will exercise in some way. Whether through jumping, climbing, giving chase or scratching, cats like to take good care of themselves as well as keep in good condition by regular grooming. This attention to self is important, both for fitness levels and general welfare.

Just as cats look after themselves well, we too can help ourselves by taking regular and appropriate exercise. Although this has to be governed by age and general condition – and here expert guidance can be of great help – keeping active can leave us feeling and looking better, as well as allowing us to do that much more. Even a brisk walk or a discipline such as yoga, Pilates or t'ai chi can make a noticeable difference.

Be like the cat: look after yourself and enjoy the benefits.

Listening Out

'The cat has a nervous ear,
that turns this way and that.
And what the cat may hear,
is known but to the cat.'

DAVID MORTON

Many cats, and certainly young ones, have excellent hearing. As hunters, their hearing allows them to detect the most minute rustlings and squeaks, and from which direction and distance they have come. Even when dozing, if a cat hears a noise that is of interest, it can be alert in moments. Cats have incredibly sensitive hearing, which is why they can be unsettled in noisy environments.

Although our own hearing is not as acute as that of cats, it's important that we value this most important sense. Modern-day life bombards us with sounds. If not traffic, it could be machines, continual chatter, media and much else besides, and in time our hearing can suffer as a consequence.

While it can be difficult to escape noise, it can often be good to find somewhere quiet to listen … just listen. Listen to the sounds of nature: the birds (always of interest to cats), water trickling from a fountain, lashing rain or, indeed, rare as it is, pure silence.

Cats' ears are very attuned to the sounds around them. Just as they try to avoid noisy environments, so it can be good for us to sometimes escape the noise around us and enjoy some quiet.

And to listen.

Just listen and appreciate.

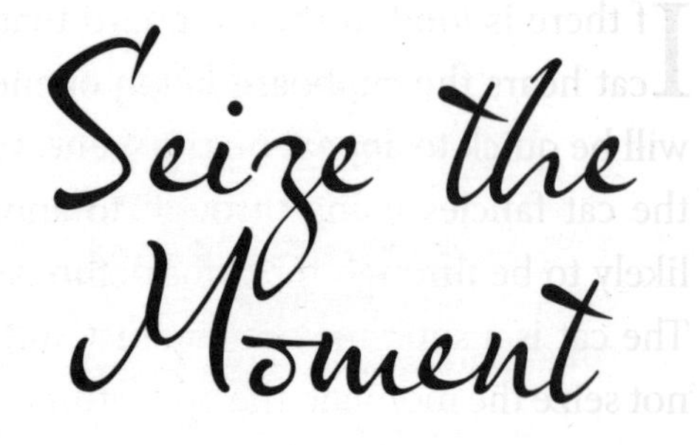

'The whir of a can opener … will send even the most deeply dozing cat bounding into the kitchen.'

BARBARA L. DIAMOND

If there is food in the cupboard that the cat likes and the cat hears the cupboard being opened, you can be sure it will be quick to appear on the scene. Or if a door opens and the cat fancies going through to another room, the cat is likely to be through the door before you realise it has gone. The cat is a supreme opportunist and knows that if it does not seize the moment, the opportunity could quickly be lost.

As with our own lives, opportunities do not last for long. They need to be taken and acted upon when they appear. Be like the cat and make the most of chances *while they are there*. That way you will also stand a better chance of benefiting from potential gains.

Sweet Dreams

'[A cat is] a dreamer whose philosophy is to
sleep and let sleep.'

SAKI

A cat enjoys its sleep, and while asleep it will often dream. This can be apparent in a sudden twitch of the whiskers, a movement of the paws, a murmur or chattering. Could it be that the cat is reliving some of the images of the day or acting through an imaginary situation, just like many of us do while asleep?

Dreaming has its value, particularly in allowing us to process what has been going on in our lives, to work through emotions, to imagine scenarios and what could be. Some – particularly the creatively inclined – find that ideas and inspiration come to them while in a dream, ready to be acted upon the next day.

Cats need their sleep, and dreaming is an important and necessary part of the process – some even say that cats lead a dream life. It is good for them, and it's good for us too. Rather than ignore or forget our dreams, we should remember that what goes on in our brains while we sleep can be restorative, benefiting us emotionally and psychologically, and helping us solve problems and explore our imagination.

Dreams, both for us and for cats, can indeed be sweet.

Stress

'Cats don't like change without their consent.'

ROGER A. CARAS

Cats do not like stress, and when stressed they send out signals that let us know something is wrong. These could be avoiding contact with others, altering their eating habits, excessive grooming (or, for some, ceasing to groom altogether) or becoming restless and more vocal. Stress manifests itself in many ways, but with the cat we can be left in little doubt that it is not itself and needs some help.

In our own lives we often experience times of stress and also send out tell-tale signs. Bizarrely, one of them is a bit cat-like. When scared, a cat will raise its fur to make it appear larger. For us, hair rising on the back of our neck can lead to us rubbing this area to help soothe away stress. Similarly, tightly interlacing fingers, wrapping feet around the legs of a chair or covering the mouth when speaking all signify unease.

Cats visibly – and sometimes vocally – let us know when they are stressed and, in a way, reach out for help, but there are times when we will try to conceal our anxiety, and it's often to our detriment. But by recognising and becoming more aware of signs of stress that we and others give, we can better understand and more effectively address the problem.

Stress affects cats and us, and in both cases we need to be noticing *and* responding.

Private Moments

'With a yawn
and a stretch,
up it gets
and off it goes.
The cat,
into a world of its own.'

NEIL SOMERVILLE

There are times when a cat will disappear and leave its owner puzzled as to where it has gone. Sometimes, on investigation, it may be found curled up under a bed or in a secluded spot in the garden. But the cat simply felt that it needed time to be alone.

Such moments are good for cats … and for us too.

We all need private moments. Time to be alone, to be still or just to get on without being disturbed.

If you feel you need such a moment and there are others around, just say, 'I need a few moments to myself,' and they will understand. Or, if you think another needs time to themselves, respect this and leave them be.

We all need our own private time, and like the cat, who hides somewhere it can just be, we too need to make sure we have times when we can be alone and be in our own private world.

Going for Best

'A plate is distasteful to a cat, a newspaper still worse; they like to eat sticky pieces of meat sitting on a cushioned chair or a nice Persian rug.'

MARGARET BENSON

Cats like the best. If given a choice of different cat food, they will usually opt for the more expensive brand, which often has tastier content. In the home, cats will also seek out what is most desirable. Whether it's a plump cushion, a particular bed or an especially cosy spot, they know what they like ... and will often seek out the finer things in life.

Just as cats can be selective, there are times when we too should follow their example and go for the finest option. Whether it's for the luxury and reliability that quality offers or additional pleasures and comforts, enjoying the best can be a special treat and a just reward for the many things we do.

Be like a cat and, when you can, enjoy some quality.

You deserve it!

Being Attentive

'That got me a look so intense I was unable to interpret it – like the way cats sometimes fix on you. What they mean by the look is completely beyond understanding; but it's meant for you, you alone.'

DONALD JAMES

When a cat looks at us with its large, longing eyes and then rubs against us, we can be in no doubt that at that very moment we are at the centre of the cat's attention. To such looks and concentrated effort, we respond accordingly.

Likewise, for us, giving our full attention is so very important in our dealings with others. With so many distractions in modern-day life, our minds can often be elsewhere and our chain of thought frequently interrupted. As a consequence, we miss out on things happening in the present, and our relationships and conversations can suffer.

Here we can learn from the cat. By giving our full attention, we can be more effective, enhance our relationships and have a much better understanding of situations.

Beware the Stare

'The two cats never fought, physically. They fought great duels with their eyes.'

DORIS LESSING

In the cat world, the stare sends powerful messages. In particular, when two rivals meet, they will try to outstare each other, and the one who closes its eyes or looks away first is the one signifying surrender.

Bizarrely, when a cat enters a room full of people, often it will gravitate towards those who do not necessarily like cats. The reason being that these are the people who do not tend to stare at the cat or try to ignore it, while cat enthusiasts usually follow its every move. Cats do not like being watched; it makes them wary and uncomfortable. So when you find yourself doing this, some blinking can be especially reassuring for the cat, who, more than likely, will close its eyes in response.

But just as cats become uneasy when being stared at, this also applies to us. In many cultures it is considered rude to stare, and although there are exceptions, including lovers gazing fondly at each other or those locked in debate, we need to remain mindful of how long we look at each other and how a prolonged look can cause unease.

Cats regard staring as a threat and we too should be aware of the awkwardness this can cause. As the cat finds, blinking and averting your gaze from time to time can break the tension and be reassuring as well as help in your dealings with others.

You Never Know

'A sleeping cat is ever alert.'

FRED SCHWAB

Cats may be in a sleepy state but they keep one eye open. They do not like to miss anything. And if an opportunity comes along or a door opens, they can be quick to take advantage. Similarly, the cat's ear may twitch when it hears a sound that interests or threatens them.

Keeping an eye or ear open is particularly important. In our own lives, we need to be alert to possibilities and chance. We need to be ready if something of interest or value comes along.

To get on, and potentially profit from situations, be like the cat and stay alert. Who knows what you may see or hear, but sometimes it could be very much to your advantage, so keep an eye or ear open.

You never know.

Sensing

'For the cat is cryptic, and close to strange things which men cannot see.'

H. P. LOVECRAFT

With their heightened senses, cats are very aware of the conditions around them; they are even able to detect changes in the atmosphere. It is for this reason that cats have had a long association with the sea, with sailors taking them on board ship to warn of bad weather ahead. Similarly, there have been many instances when cats have sensed other kinds of danger, alerting their owners to impending earthquakes and potential hazards in the home, especially fire risks such as overheating appliances.

While our own hearing, sight and smell may be no match for the cat's, we should take heed of our own senses. If we feel something is not right or an inner voice is telling us to be wary, we should take note.

Cats are very responsive to what is going on and use their heightened senses well. By listening to our own intuition, we can often be guided to the right actions and alerted to those that may not be in our best interests.

Like the cat, be aware and listen to your inner voice. It can be a good indicator.

Moving On

'A cute expression and a purr, and the latest shredding incident is forgiven.'

ANONYMOUS

In many a home a cat has done wrong. My own cat, Roly, once knocked over an attractive vase, something which upset my mother and concerned me. And I remember Roly slinking off while my mother and I cleared up the debris. For the rest of the day, he kept a low profile.

The next day Roly was back to being his adventurous self. He had moved on, although he never again ventured onto the shelf where the accident had occurred.

In our own lives, accidents and mishaps occur, but like Roly and other cats who have done wrong, we should remember that what has happened has happened. Lessons can be learnt and new starts can be made.

Although we may regret some of the wrongs we have done, mistakes we have made and upsets we have caused, we must move on – wiser and sometimes bruised, but move on we must.

'Any household with at least one feline member has no need for an alarm clock.'

LOUISE A. BELCHER

It has often been said that 'cats are creatures of habit', and they do welcome routine and pattern to their day. They are well aware of the time their owners get up – and often rouse them, should they oversleep. They even seem to know when their owners leave for work and when they return, usually getting in position moments before to welcome them back. They have a good internal clock.

Cats also know – and look forward to – feeding time, as well as the time when their owners generally sit down to give them a bit of extra fuss.

Routine gives structure and helps cats – and us – prepare for what lies ahead. Having a routine also helps us to be more efficient and better prepared, sometimes freeing up that most precious of commodities: time.

Cats like routine and there is much to be gained by giving thought to your own, as well as the best and most efficient ways to organise your daily activities.

Enthusiasm

'It is, of course, totally pointless to call a cat when it is intent on the chase. They are deaf to the interruptive nonsense of humans. They are on cat business, totally serious and involved.'

JOHN D. MACDONALD

If given a new toy, a cat is likely to regard it with considerable interest, pawing it, smelling it and seeing if it makes any noise. As it does so, it will set about it with considerable intensity – and often to our amusement.

But there is something more.

A cat's passion, focus and energy draws us in ... We watch what it does with fascination.

This is what passion and enthusiasm does.

When you show enthusiasm, it is often conveyed to others and draws them in. Enthusiasm *is* contagious, and just as we take pleasure in watching a cat's enthusiasm, so the enthusiasm we ourselves generate can be a joy, a spur and an inspiration to others.

As you set about your activities, inject some enthusiasm. It can make a lot of difference and, as many a cat has shown, gets others interested and possibly involved.

Enthusiasm is like a magnet. It has great pulling power.

On Making It Clear

'Cats ask plainly for what they want.'

WALTER SAVAGE LANDOR

Many owners find that their cat has a different meow for different situations, which they soon get to know. It could be a 'let me out' or 'let me in' meow, an affectionate 'give me attention' meow, an annoyed meow or a 'please feed me' meow; each nuanced sound is an effective way for a cat to tell us what they want. Despite such a limited repertoire, it is remarkably effective.

This ability to draw someone's attention to what you want is important, especially with others often so busy. With cats there is no beating around the bush. One particular meow is usually all it takes, and for us it can prevent misunderstanding to be direct and make sure others know *precisely* what it is we want.

Just as cats draw our attention to their needs – and make sure we know exactly what they are – so it is to our advantage to be precise and clear. That way you will stand a far better chance of getting what you want – something a wise meowing cat knows all too well.

Standing Up

'Some people say man is the most dangerous animal on the planet. Obviously those people have never met an angry cat.'

LILLIAN JOHNSON

Most cats avoid confrontation, but there will be times when they take a stand and defend what is theirs, most often territory and space.

And just as cats stand up to each other when circumstances dictate, there are times when we too need to stand firm and defend what is precious to us, including our principles and beliefs. We may not like or wish for such scenes or unpleasantness, but if something is vital to our interests, others need to be made aware of our position.

Cats are fairly tolerant creatures, but they also realise the importance of taking a stand *and being heard.* Despite their small size, they are no walkovers and often succeed in getting their way. When something important is at stake, we too need to make sure that our position and feelings are clear to others.

When the situation calls for it, the cat is a force to be reckoned with. So are we.

Win-win

'Nobody who is not prepared to spoil cats will get from them the reward they are able to give to those who do spoil them.'

SIR COMPTON MACKENZIE

Cats bring joy to many a home. And many cats enjoy being in our homes, valuing the security and comfort it brings – as well as the good supply of food.

In return a cat offers company, which, because of its calming presence, can also be good for our health. By welcoming a cat into our home, we value the benefits it brings. The relationship is win-win. A win for the cat, as (to use a cliché) the cat always seems to know which side its bread is buttered on, and a win for us.

This principle holds good for many things. As we go through life and enter commitments, striving for a win-win situation offers advantages for all – for you *and* the other party. The cat, in its relations with us, knows that it benefits from it, and we can strive for the same in other strands of our life.

'Win-win' was one of the habits Stephen R. Covey wrote about in *The 7 Habits of Highly Effective People*. As he noted, 'Win-win sees life as a cooperative arena,' and one in which 'we both get to eat the pie, and it tastes pretty darn good!'

I am sure the cat, an expert in win-win, would second that.

Poised to Do Well

'Cats never strike a pose that isn't photogenic.'

LILLIAN JACKSON BRAUN

When venturing out, many a cat moves with dignity and poise. With its head held high, its long body and trailing tail, the cat makes an attractive figure. And in its slow, deliberate movements, it carries itself with a certain grace. In so many ways, cats are blessed with style and are a joy to watch.

In this we should learn from the cat. The way we move is important and it's worth giving thought to our poise. Slouching or just shuffling along can suggest a 'couldn't care less' attitude, whereas by walking with awareness and your head held high, you exude a certain dignity and confidence, and look much more impressive.

The cat has poise, and taking note of your own posture and walking well, not only helps you look good, but feel good too.

Self-sufficiency

'A cat keeps her claws sharp because she knows a purr might not be enough.'

ANONYMOUS

Although most domestic cats are well fed and have no need to hunt, they are often keen to maintain their hunting skills. That way they can be self-sufficient and fend for themselves should the need arise.

Although our needs are very different to those of a cat, there is much value in maintaining and acquiring abilities that can help us both now *and* in the future. Whether you decide to learn a new skill that could be of practical use or may help your income and prospects, or you try to rely more on yourself and your abilities, you could find what you do a satisfying investment in yourself.

Within yourself are the riches of your tomorrow.

Like the cat, it pays to keep your abilities nicely honed.

Knowing Where

'Cats do not go for a walk to get somewhere but to explore.'

SIDNEY DENHAM

Cats like to build up a mental map of their surroundings and territory. They like to know where things are, where's best to go (including the most fruitful hunting areas) and the places to avoid. And their knowledge of their immediate vicinity serves them well. On most days cats will make a point of checking their area and investigating anything that may have changed.

This sense of knowing where things are can be especially useful for us. In particular, in the home, knowing where certain things are – and having a place for keeping those things – allows us to be tidier, better organised and more efficient. Just as cats like to know where things are, by doing this in our own home (and place of work) we stand to benefit – not least by saving wasted time spent looking for things.

Be like a cat: make a habit of knowing where.

Look Before You Leap

'Unlike us, cats never outgrow their delight in cat capacities, nor do they settle finally for limitations. Cats, I think, live out their lives fulfilling their expectations.'

IRVING TOWNSEND

When a cat prepares to jump it focuses on where it intends to land, sizing up distance and what is needed to reach its target. Even if that target is some distance away and requires a final scramble, the cat is usually successful in getting to where it wants to go. It is certainly a good 'sizer-upper', and this is something we can learn from them too.

If there is an aim or objective you want to reach, spend some time sizing up what is needed to get to it. And, rather than jump into projects or activities blindly, prepare yourself and consider what is involved. That way you will stand a far greater chance of reaching your target. Be like the cat: size up objectives and know *where* you are heading.

Getting to Know You

'There are no ordinary cats.'

COLETTE

Cats are very adept at revealing their true selves. When meeting a cat for the first time, you will often gain a good idea of what the cat is like. In particular, whether it is shy, affectionate, playful, a grumpy 'leave me be' sort or if it's at ease in human company or not. Because it lets its nature be known, we know how best to respond to it.

This transparency can also be helpful in our dealings with others. If you are open and upfront, those you meet will be better able to gauge and respond to you, and your relationships and conversations will often benefit as a consequence. For instance, when being introduced to someone new or striking up an initial conversation, it could be helpful to mention a key interest of yours or an unusual fact about yourself, as well as showing an interest in the other person and what they are saying. By doing so, it can help build a connection and establish a rapport.

Cats make it easy for us to know how to treat them by letting us know what they do and don't like. And by opening up more to the people you meet, you'll find it will generally help to make your dealings with them more interesting and ultimately more rewarding.

Contentment

'Purring would seem to be, in her case, an automatic safety-valve device for dealing with happiness overflow.'

MONICA EDWARDS

A cat's purr is often a sign of contentment. Whether it's being stroked or groomed or meeting someone it likes – be it another cat or a human – a purr is an indication of the cat's pleasure. Cats can also purr as a source of comfort, with some purring when unwell.

Just as cats use purring in a beneficial way, we too can follow suit, although rather than purr, we could hum, whistle or sing. As cats have found with purring, our own contented sounds can help lift the spirits. A hum, whistle or song can put you in a good mood, and for those who may overhear you, it can help give them a boost too (provided you are reasonably tuneful, that is!).

A hum or, for a cat, a purr, can certainly do good.

'No amount of time can erase the memory of a good cat, and no amount of masking tape can ever totally remove his fur from your couch.'

LEO DWORKEN

Being so territorial, cats are experts at leaving their mark. This is often done through scent marking or more visual methods, such as scratching.

Just as making their mark is important for cats, as another means by which they can communicate, this also applies to us. Our mark is our signature, and it's something to take pride in. It represents *you*, and by making it clear and perhaps by adding a distinguishing feature – maybe a squiggle, a line or making a certain letter prominent – it can help give you a stronger identity.

As with cats, the mark we make is important, especially in the impression it leaves and the way it represents us.

Take pride in your signature and how *you* make your mark.

Never Too Late

'Cats have a sense of humour, as is shown in their extreme love of play. A middle-aged cat will often play as unreservedly as a kitten, though he knows perfectly well it is only a game.'

WILLIAM LYON PHELPS

Kittens like to play – and it does them good, providing moments of fun, helping them explore their abilities and encouraging social interaction. However, as kittens grow and reach about five months, their tendency to play lessens.

But it is never too late.

If a more mature cat has the opportunity to play, or ideally if you play games with your cat, it will often be keen and ready. And there is a lesson to be learned here.

In previous years you may have thought about taking a skill further, travelling to a particular destination, writing a book or achieving some other goal. If so, why not revisit the idea and see what is possible?

As cats regularly show, it is never too late to rekindle the joys and pleasures of playing, or to reignite certain hopes, plans, hobbies and activities that we may have previously thought about.

It is never too late … and by giving things a go, you never know what may come as a result.

Body Clock

'The cat has always been associated with the moon. Like the moon it comes to life at night, escaping from humanity and wandering over housetops with its eyes beaming out through the darkness.'

PATRICIA DALE-GREEN

For much of the day a cat may appear to take life at a leisurely pace, sleeping, catnapping and occasionally wandering off to inspect its territory. But at dawn and dusk many a cat becomes more active. This is a time for prowling, foraging, chasing, hunting and sometimes getting into mischief. According to their body clocks, this is a time for action.

Just as cats make the most of the times that feel right for them, you too should take note of your own body clock and, when possible, keep your most important tasks for when you feel at your optimum. For some this can be first thing in the morning, for others late at night, but like the cat, listen and respond to your own body clock and make the best of the times that suit you.

Final Thoughts

'By associating with the cat,
one only risks becoming richer.'

COLETTE

Cats are celebrated in all sorts of ways. In Ancient Egypt there was Bastet, a sacred cat goddess who was much revered, and throughout the centuries feline skills in pest control have been greatly valued. Although it's not certain when cats first became domesticated – some have estimated it could be about 9,500 years ago – they have long been a companion to mankind, and today their presence is everywhere. Not only do they grace many homes, they have also found their way into literature, art and music, and onto stage and screen. Even in the road, cat's eyes (inspired by a cat) are a visible and important presence. Cats are all around ... often bringing joy as well as companionship.

As this book draws to a close, take a moment to think about what the cat – possibly *your* cat – means to you and has brought to your life. Think of the special moments shared, perhaps a time when a cat looked at you, you blinked and the cat closed its eyes for a moment in response, or when a cat's antics especially amused you. Or perhaps, when you were feeling unwell, a cat curled up with you, providing comfort.

Cats can give us special moments that linger long in the mind.

Treasure those moments. They are all part of the joy of spending time with cats, and by recalling such times, we often find that they have put a smile on our face or brightened us up in some way. Such is the specialness of cats … their unique contribution and wisdom.

List of Lessons in Alphabetical Order

Adapt
At Ease
Awareness
Being Attentive
Beware the Stare
Body Clock
Buying Time
Contentment
Courage
Decision Time
Discover
Downtime
Enthusiasm
Exercise
Favourite Places
Friendship
Getting to Know You
Going for Best
Gratitude
The Importance of Giving
Keeping Your Head Down
Knowing Where
Language of the Eyes
Letting Off Steam
Listening Out
Look Before You Leap
Looking Good
Make Your Mark
Making Do
Moving On
Never Too Late
Nine Lives
On Making It Clear
Persistence
Poised to Do Well
Private Moments
Quiet Observation
Reflecting
Responsibility
Reward
Routine
Scent
Seize the Moment
Self-sufficiency
Sensing
Sleep
Special Greetings
Standing Up
Stress
Stretching
Surroundings
Sweet Dreams
Taking Care
Taking Stock
Tell-tail Signs
Value the Now
Venture Hopefully
Win-win
Worth the Wait
You Never Know

About the Author

At the age of four the author was taken to see some kittens playing on a heap of coal and allowed to choose one. It was a fluffy black-and-white one that was rolling about that caught his eye. He named him Roly. It was then that Neil's love of cats began, enjoying their personality, company and specialness. History was to repeat itself later when Neil's children got their own black-and-white kitten, Smartie, who became another wonderful friend to the family. In writing *Cat Wisdom* he has been able to draw on his observations of cats and share in the delights and wisdom offered by our furry feline friends.

Neil is also a great believer in positive thinking, as well as having an interest in Eastern beliefs and traditions. For thirty years he wrote the international bestseller *Your Chinese Horoscope*, and he is also the author of *Your Chinese Horoscope for Each and Every Year* and *The Answers.*

In addition to cats, his other interests include travelling and writing haikus.

Neil lives in Berkshire, England, with his wife, and has two adult children and an adopted cat, Lily.

Acknowledgements

My own and other people's cats have helped me so much in the compilation of this book, thanks are also due to my family – Ros, Richard and Emily – for their thoughts and cat reminiscences. A special mention too for my parents, Peggy and Don, who encouraged my love of cats, to my editor at HarperCollins, Carolyn Thorne, for her great enthusiasm, and to Barbara Smith, Joan Moules, Barbara Booker and Rene Scott for their faith and support. To David Finnerty, again it was good to talk cat matters. To you all: thank you – as well as to you, the reader, for your interest in *Cat Wisdom*